SELF-ORGANIZATION AND MEDIATED TRANSIENCE IN PLANT COMMUNITIES

What are the rules?

"Alle Gestalten sind ähnlich, und keine gleichetder andern;
und so deutet das Chor auf ein geheimes Gesetz, ..."

Johann Wolfgang von Goethe (1798): "Die Metamorphose der Pflanzen "

http://www.udoklinger.de/Deutsch/Goethe/quelle.htm

In free translation:

"All [plant] forms are similar, but none like any other;
and so the assembly interprets a hidden law ..."

I owe Márta my sincerest thanks for
having graciously tolerated yet another
period of my preoccupation with research projects.

In brief: what is inside the Book?

A novel, signal theoretical solution is sketched out for the ecological problem of how to identify and quantitatively express the assembly rules of plant communities. A case study for testing the solution leads to the astonishing conclusion that the phylogenetic signal outperforms the current environmental signal in intensity close to 7 to 1. This indicates high stability and low inclination to environment mediated transience in the community.

Keywords: Community assembly/disassembly, Coquihalla floodplain, Dendrogram, Diversity moderation, Governance rules, Environmental mediation, Environmental signal, Evolution, Floodplain, Hierarchical relevé, Partial correlation, Partial variance, Plants, Phylogenetic signal, Phylogenetic tree, Scaling functions, Signal isolation, Vegetation

SELF-ORGANIZATION AND MEDIATED TRANSIENCE IN PLANT COMMUNITIES

What are the rules?

László Orlóci, FRSC

Visiting Professor, Laboratory of Plant Quantitative Ecology, Department of Ecology, Universidade Federal do Rio Grande do Sul, Porto Alegre, Brazil

London, Canada - 2012

Refer to this book as

Orlóci, L. 2012. Self-organisation and Mediated Transience in Plant Communities. Scada G26Publishing, London. Online Edition: CreateSpace https://createspace.com/3585127

Look for

Orlóci, L. 2011. Statistical Ecology. The quantitative exploration of nature to reveal the unexpected. Scada Publishing, London. Online Edition: CreateSpace https://createspace.com/3476529

Orlóci. L. 2011. Problem flexible computing in statistical ecology. Scada Publishing, London. Online Edition: CreateSpace https://www.createspace.com/3574792

Orlóci, L. 2012. Statistical multiscaling in dynamic ecology. Probing the long-term vegetation process for patterns of parameter oscillations. Scada Publishing, London. Online Edition: CreateSpace https://createspace.com/3830594

ISBN 978-1461028222

ACKNOWLEDGEMENTS

The credit for expert data goes entirely to Ms. Márta Mihály DFE. The basics signal scaling and isolation, and the programming code was completed during my tenure as Visiting Professor of Statistical Ecology in the Department of Ecology, Laboratory of Quantitative Ecology, Universidade Federal do Rio Grande do Sul, Porto Alegre, RS Brazil. I owe to Prof. Dr. Valério De Patta Pillar and esteemed colleagues in the Department my sincerest thanks for unabated warm reception on my successive visits and for enlightening conversations. Thanks are extended to academician Dr. János Podani, Dr. Marcela Pinillos and Prof. Mathew Mukkattu for comments and perplexing questions.

CONTENTS

RETROSPECT

We begin with a brief account of literary events from the longer past which we consider necessary to dispel the student reader's possible impression that the Book's central topic, the search for rules of plant community governance, heralds a novel undertaking by us or by others of our contemporaries. The search for intrinsic rules has in fact been the principle incentive of the development of Ecology as a science over now considerably more than two centuries.

The early period is noted for the publication of the elegiac poem "Die Metamorphose der Pflanzen" in 1798. The poem's author, the naturalist Johann Wolfgang von Goethe, speaks of a quire of plant forms which interprets a hidden law purely by the similarities of the individuals in an otherwise diverse quire membership. Just by changing the identifiers from 'plant form' to 'plant individual' -- which is a legitimate thing for us to do, since no two of the quire members are considered exactly alike -- the quire now as a plant community is interpreting its own hidden law referenced to member plants' similarity and diversity.

Speaking of a hidden law, and linking it to the similarity of members in an otherwise diverse natural collection of plants, must have appeared at the time of revolutions as itself a revolutionary science. The more so, since science's understanding of the plant community concept at that time have had yet to progress beyond the aesthetic plant geographic view, which Sukopp (1987) finds in Humboldt and Bonplan's "Essaie sur la Géographie des Plantes" (1805).

Goethe's ideas on the morphology and functioning of plants, and his clearly expressed notion of plant assemblages discharging functions like a quire in the presence of a hidden law, did not go unnoticed among the 19th Century naturalists. It may not be a far fledged proposition that Goethe's ideas are precursor to Kerner's doctrine of plant community development. Kerner does in fact quotes Goethe in his seminal book "Das Pflanzenleben der Donauländer" (1863).

Kernerian plant community development is a site related point process in time. It involves compositional transitions in situ mediated by action-reaction feedback. Ecologists identify this mechanism -- with or without an awareness of Kerner's work -- as facilitation which can generate community transience. The central core of Kerner's doctrine is thus the proposition of a mediated tendency in plant community functioning to change the environment at the potential cost of its own transience.

How did Kerner come to the idea of temporal compositional transitions undertaken by the plant community in situ? His approach is a case of space-for-time substitution[1]. It could go like this:

---

[1] Wildi and Schütz (2000) describe a complex case. They take fractional time series from permanent plots and splice these into a single chain. This chain is used as surrogate in the "prediction" of succession over the time span of centuries. The permanence of the same kind of climate is assumed.

1. General reconnaissance of highly dynamic alluvial sites[2] reveals coincidental patterns of plant community types and nominal substrate age.

2. Space (the pattern of community types) linked to time (sediment age), the conclusion of plant community development by way of temporal compositional transitions in situ is in hands.

We should mention as corollary to the approach, a detracts from its practical value. Namely, the developmental series is defined for community types which happened to be observed in the studied site. Community types potentially present but not observed are left out.

---

[2] We refer to Kerner's Achen Lake example from Tirol.

SYNOPSIS

The book leaves it to the more ambitious reader to feel the gap left by what has been said in retrospect and what is going to be discussed in detail. The book focuses on two complex functionalities: *biological self-regulation* and *current environment mediated community transience*. Much knowledge accumulated about such functionalities since Kerner's (1863) time and it is absolutely clear that in the natural plant community no complex functionality can be entirely ruled by the law of chance. Yet, chance in the manner of random events happening is important. We applies a novel approach to show this and to identify high-level governance rules other than the rule of chance.

At the bases of our approach is signal superposition as seen in the model $TS=PGS+CES+e$. Accordingly, diagnostic 'signals' having to do with phylogeny and current environmental mediation, symbolically *PGS* and *CES,* are superposed in convolution with signal *e* linked to random events. As such *e* confounds *PGS* and *CES,* and through this, it controls the level of chance oscillations in biological self regulation and current environment mediated community transience.

A unique property of the signal sources is the convolution of their total effect (signal *TS*). The book derives scaling functions and signal isolation techniques by reasoning from two very realistic assumptions: the total signal's *(TS)* linkage to community's diversity state; and the independence of the phylogenetic signal *(PGS)* from the current environmental signal *(CES)* .

The book's contents begin with a set of propositions. These define the context both in the problems' formulation and in the conceptualization of the solution. Theoretical foundations for signal isolation and scaling, presentation of techniques, and a comprehensive discussion of results from a real example make up the book's main body. Key literature references, an index, and appendices of the numerical results round out the contents.

PROPOSITIONS

From the results presented in the book we have drawn specific generalizations about the assembly/disassembly rules of plant communities. We note that we consider point processes in a landscape context, therefore by definition community assembly and disassembly must occur in tandem. A consequence of this is that in reality assembly and disassembly in Nature are the two sides of the same process called "succession".

The propositions:

1. Community assembly/disassembly is a complex-systems functionality. It is governed by system-level natural rules for which the phylogenetic signal *PGS,* the current environmental signal *CES,* and *e,* a signal associated with random events, are diagnostic.

2. Signals *PGS, CES,* and *e* occur in convolution for which the proposed model is *TS=PGS+CES+e.* The model implies linear superposition of the signals. The model does not have nor does it need an interaction term.

3. The intensity of *PGS* is indicative of the intensity of biological self-organization in the community forced by the interactions of community elements.

4. The intensity of *CES* is indicative of the level of current environment mediated transience in the plant community.

5. Consistent with the linear superposition model, two additional generalizations are suggested:

a. When *PGS* is the dominant signal (relative to *CES*) and the *e* signal is intense, highly turbulent interactions and commensurably high community stability are expected, the near-climax state of succession.

b. When *CES* is the dominant signal (relative to *PGS*) and the *e* signal is weak, transiency is intense and community instability is elevated.

In summary, while the *PGS* signal is symptomatic of the governance of self-organization in the community, the *CES* signal is symptomatic of controls in environmental mediation of community transiency.

FURTHER ON SIGNAL SOURCES

We already identified the signals, now we consider the sources in somewhat more detail:

1. *Signal PGS.* The source of *PGS* is the *historic phylogenetic process.* This process is responsible for speciation and rather directly for the enrichment of the global flora. We stress that the basic taxonomic unit in the present study -- unlike in our work earlier, which form the conceptual basis of signal isolation in this book, and in which we use taxa that were character set types (Orlóci 1991) -- at this time the taxa we use are Mendelian species.

Why do we need Mendelian species? Simply stated, the definition of *PGS* and its isolation requires the taxa to be meaningful evolutionary units. Further in signal isolation and scaling we rely on a *taxonomic dendrogram* for proxy of the *true phylogenetic tree*[3]. On this topic we refer the reader to text

---

[3] Following text is quoted from
http://en.wikipedia.org/wiki/Phylogenetic_tree :

"A phylogenetic tree or evolutionary tree is a branching diagram or tree showing the inferred evolutionary relationships among various biological species or other entities based upon similarities and differences in their physical and/or genetic characteristics. The taxa joined together in the tree are implied to have descended from a common ancestor. In a rooted phylogenetic tree, each node with descendants represents the

which presents syntheses of the evolutionary principles (Huxley 1942, Dobzhansky 1937, Stebbins 1950, Mayr 2002, Mayr and Provine 1998, Fenzenstein 2004, Podani 2003, 2010, Schuh et al. 2009). [4]

2. *Signal CES.* The source is *current environmental mediation.* The signal has to do with species trait assortment over the landscape's points into environment specific plant communities. Two aspects of this need further clarification. 'Current' is emphasizing that the environmental mediation we have in focus is in the 'now' *vis-à-vis* the process of 'speciation' which as it applies in our problem is in the long past. Considering that we are dealing with environmental mediation as a point process, the two processes highlighted by *PGS* and *CES* meet in the present where the species traits for which evolution is responsible are being assorted by the other process, environmental mediation. The sources meet at just one point in time, therefore *PGS* and *CES* justifiably considered practically independent in the signal isolation problem.

3. *Uncontrollable random variation.* This is the source for *e*, encompassing all random behaviour in the plant community and environment. This source makes *TS* fuzzy and its percep-

---

inferred most recent common ancestor of the descendants, and the edge lengths in some trees may be interpreted as time estimates. Each node is called a taxonomic unit. Internal nodes are generally called hypothetical taxonomic units (HTUs) as they cannot be directly observed. Trees are useful in fields of biology such as systematics and comparative phylogenetics."

[4] The importation of evolutionary principles into plant taxonomy is an on-going process ever since the work of G. Bentham and J.D. Hooker (1862-1863), the first to break with the numerical system and to adopt Darwianian principles in plant taxonomy. The rules formalised based on a synthesis of all evolutionary theories (see Stebbins 1950) and reached mathematical definitions in Cladistics (see Henning 1966, Singh 2004, Podani 2003, 2010) by applying the tools of numerical taxonomy.

tion as a directed (non-random) signal stream proportionately more difficult.

An important property of the sources is the superposition of their effects into a common forcing effect. The total forcing effect modulates the amplitude of species trait diversity in the community. Significantly, the diversity connection hands us scaling functions and an effective signal isolation algorithm. The framing of signal isolation in these terms is completely unlike in early attempts which Revell et al. (2008) categorically rejected.

PILLARS OF SIGNAL ANALYSIS

It flows directly from what has already been said that there have to be access to several attributes to make possible signal isolation and scaling. What are these?--

1. *A species list amended with abundance estimates.* Abundance refers to any of the variables applied in Ecology (Orlóci 2010), ranging from cover/abundance (C/A) to mass or volume based biomass.

2. *A long and crisply formed environmental gradient.* The user defines the gradient and based on careful planning populates it through its entire length with real sampling units, such as the ecologist's quadrats or sample plots. Gradient length is relative to the ground scale of the vegetation and environmental pattern (Greig-Smith 1952). The variance of the dominant gradient variable U is a measure of the gradients crispness.

3. *An evolutionary plant taxonomic system.* It has to be assumed that species identification is based on a taxonomic system. Only when we use an evolutionary taxonomic system can we identify Mendelian species traits and construct a *taxonomic dendrogram* which may qualify as proxy for the *true phylogenetic tree.*

4. *An algorithm for signal isolation.* Our method for this is in fact relatively straight forward. Variance partitions are involved under constraints similar to what has already been discussed in earlier works (Orlóci and Orlóci 1985, Orlóci 1991, Pillar and Orlóci 1993).

FURTHER ON LINEAR SUPERPOSITION

The model $TS=PGS+CES+e$ is a conceptualisation of how the signals stack up. Reader may be wondering already why the model does not include at least one product term. Not having such a term implies the assumption of the linear independence of the signals. We already explained that signal independence should not be seen as a far fledged expectation if one considers the fact of time separation of the signal sources for PGS and CES.

We mentioned also that variation along the environmental gradient is in the present, environmentally mediated assortment of species traits is very much 'current', but the evolutionary origin of the species traits that make up the local flora is far history (see section above). Questions could of course be posed to the effect that since different traits of the same species may populate different habitats, the adaptive process which produced the traits had to be linked to habitat conditions modelled on the existing ones. In other words, speciation must be in progress. These are all true, but in all likelihood the traits being currently recorded long existed and must have been the object of environmental assortment cycles into progressively assembled and disassembled communities, including also the ones which presently exist, if by no other forcing than the Kerner feedback loop of facilitation. Therefore, the process that produced the traits and the process which mediated current assortments over the landscape points have to be considered independent in time.

Yet another question could bring up the fact that the species traits subjected to current environmentally mediated selection and assortments are the products of the process which gave rise to *PGS,* and for that reason the existing species traits establish a link between *PGS* and *CES.* The point here is that *PGS* and *CES* meet at one time point: the species traits which phylogeny produced in the long past are subjected to environmental mediated selection and assortment in the present. Before that singular point in time, the two processes have never met. Having said this, we are justified to regard *PGS* as being symptomatic of complex-systems type self-regulation. This is different from what *CES* is involved with, namely environmental mediation, for which a typical feedback mechanism is Kernerian facilitation and the result is transience.

PATH TO THE PAST

Each sampling unit is described by a record set called *relevé*. The most unique aspect of the relevé in our problem is the incorporation of a taxonomic dendrogram. Such a hybrid construct is called a *hierarchical relevé* (see Figure 1).

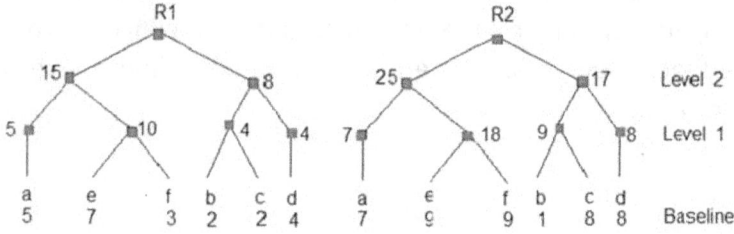

Figure 1. Two hierarchical record sets (two relevés) are shown, *R1* and *R2*. The fictitious 2+1 level dendrogram in-corporates 2 "Level 2" taxa (e.g. families), 4 "Level 1" taxa (e.g. genera), and 6 baseline taxa a,b,c,d,e,f (e.g. species). The scheme is the same in *R2* as in *R1*, only the baseline data are different. The base line (0 level) shows taxon abundance (species in this case) measured on case-specific scales (see main text). Baseline data sets in the dendrograms: *[5 7 3 2 2 4]* and *[7 9 9 1 8 8]*. Level 1 cumulants: *[5 10 4 4]* and *[7 18 9 8]*. Level 2 cumulants: *[15 8]* and *[25 17]*.

The dendrogram has *m* levels such as for example Species, Genus, Family, Class, Order, ... The mapping of each species is a trace through the nodes from bottom to top through the levels fixed by the dendrogram's phylogenetic system. Each species has an abundance estimate X in the base-line data set. Clearly when a species is found within a sampling unit, X will be greater than zero.

We re-emphasise that the hierarchical relevé amalgamates two components, the species mappings (dendrogram) and the abundance values (X), both in a seamless manner. In the discussed case, the dendrogram is based on an evolutionary taxonomic system and for that reason it should qualify as proxy for the phylogenetic tree. But can the dendrogram which only includes the species found within a vegetation stand of a local site give us more than a very small amount of information about the phylogenetic tree? Truly, it cannot. But the relevant point is the relevés ability to give an unbiased description of the vegetation. It has to be seen that the description's reliability will depend on the sampling technique. When the sampling technique is statistical (see Orlóci 2010), the information gained about the description is unbiased, and by definition, the dendrogram is a reliable proxy for the phylogenetic tree.

The accuracy of the proxy will depend on sample size and in our specific case on how sufficient is the length and crispness of the gradient, in other words, how well the transect can capture the breadth of floristic and environmental variation within the broader landscape.

We do not have to analyse the basic sampling units individually. We may choose to pool adjacent sampling units into super stands, called *metacommunities*. The larger the metacommunity, the larger is the expected number of species and the greater the statistical reliability of the dendrogram as a proxy. Since the dendrogram is constructed to accommodate all species in the sample, its reliability reaches the maximum when all sampling units are pooled into a single meta-community.

At the sampling level, the elements of **X** are species abundance values which we estimate within the sampling units (quadrat, sample plot). After pooling the sampling units, the number of species mapped into the dendrogram may increase, and the baseline data may change, but the number of elements in **X** will not. The baseline **X** and the cumulants at the dendrogram nodes are the basis of all analyses performed at any metacommunity size

SCALING THE TOTAL SIGNAL *TS*

Preliminaries

We already defined *TS* as a simple sum of terms, *PGS+CES+e* and we explained that in this model *PGS* is linked directly to community self-organization, but independent from the confounding *CES*, which we associate with current environmental mediation of the assembly/disassembly process. As explained, we associate *e* with random events. We mentioned also that *TS* is directly scalable based on its link to diversity amplitude modulation within the metacommunity.

Regarding the scaling of *TS*, we observe

$$TS_i = PGS_i + CES_i + e_i, \quad i = 0, 1, 2 \ldots, m$$

and

$$TS = TS_0 + TS_1 + TS_2 + \ldots + TS_m$$

Clearly, the series represented by the elements in

$[TS_0 \; TS_1 \; TS_2 \dots TS_m]$

defines an *m*-valued *temporal conditional (partial) diversity amplitude stream*. It should be noted that a suitable scaling function for *TS* should allow *m*-valued additive partitions. The additivity condition limits our choice to three basic scaling functions:

1. *Probability.* The probability based scaling function is a conditional probability and the series of elements in

$[TS_0 \; TS_1 \; TS_2 \dots TS_m]$

is a *temporal conditional probability stream*. Regarding the conditional probability concept we refer to text published elsewhere (e.g. Orlóci 2006).

2. *Logarithmic probability.* Rényi's (1961) *partial entropy of order one* (the partial Shannon diversity measure) is an example. The resulting series in

$[TS_0 \; TS_1 \; TS_2 \dots TS_m]$

is a *temporal partial entropy stream* (Orlóci 2006, 2010).

3. *Sum of squares.* In this case, the series in

$[TS_0 \; TS_1 \; TS_2 \dots TS_m]$

is a *temporal partial sum of squares stream* expressible in the form of a *temporal partial variance stream* (see definition below).

We chose scaling functions from the last category, the same as in the Pillar and Duarte (2010) paper. The Pillar-Duarte solution for signal isolation is based on three matrices of the metacommunity:

Matrix **P**. This matrix fixes the metacommunity's species composition. The species are given weighs according to phylogenetic trait which references the position of the species in the proxy taxonomic system.

Matrix **T** - This is the metacommunity's mean vector of functional traits of the species.

Matrix **E** (not our *E or e*). Sampling unit positions are fixed by this matrix relative to the level of environmental effect on the chosen gradient. Within the Pillar-Duarte solution, the correlation of **T** and **E,** when not trivial, supports the conjecture **E→ P →T**; that is: the environmental selection of species in plant community assembly is equivalent to selection of functional traits mediated by the phylogeny of the species.

Our analytical scheme differs in that it is formulated about the hierarchical relevé description of the metacommunity. When there are *q* metacommunities, there will be *q temporal partial variance streams* and *m spatial partial variance streams:*

$$
\mathbf{V} = \begin{bmatrix}
V_{10} & V_{20} & \cdots & V_{q0} \\
V_{11} & V_{21} & \cdots & V_{q1} \\
V_{12} & V_{22} & \cdots & V_{q2} \\
\cdots & \cdots & \cdots & \cdots \\
V_{1m} & V_{2m} & \cdots & V_{qm}
\end{bmatrix}
$$

Under the same conditions, we will have *q(q-1)/2* distinct temporal partial covariance streams.

In analytical terms, the hierarchical partition model that hands us the elements in the *spatial partial variance streams* (rows in *V*) has ecological precedent in Greig-Smith's (1952) variance based and in Orlóci's (1971) information theoretical pattern analysis.

The hierarchical relevé model as presented in this book and the temporal partial variance and covariance streams have precedent in the schemes applied by Orlóci and Orlóci (1985), Orlóci (1991, 2010), and Pillar and Orlóci (1993). The modification is in the unbalanced hierarchical embedding of taxa in the *m*-level dendrogram. The scaling functions presented in the sequel make this property of the analysis uniquely clear.

Scaling functions

The following definitions are applicable for the hierarchical relevé of any metacommunity:

1) $S_0^2 = \sum_{j=1}^{k_0} \left(X_{0j} - \bar{X}_0 \right)^2$

This is the total sum of squares; symbol $\bar{X}_0$ is the mean of k_0 values and X_{0j} is the jth element in base line **X**.

2) $S_i^2 = \sum_{j=1}^{k_i} n_{ij} \left(\frac{X_{ij}}{n_{ij}} - \bar{X}_0 \right)^2$

This is the sum of squares of the cumulants on dendrogram level *i*; symbol $\bar{X}_0$ is the mean of k_0 values in baseline **X**, X_{ij} is a cumulant at the jth node hierarchical level *i*, and k_i is the number of nodes on level *i* (see Figure 1).

3) $TS_i = S_i^2 - S_{i+1}^2$. This is the ith of *m*-values in the *temporal partial sum of squares stream* $\left[TS_0 \ TS_1 \ TS_2 \ ... \ TS_m \right]$.

TS_i is the diversity amplitude increase brought on when the *i+1* level taxa are split into the *i* level taxa, such as genera into species, family into genera, and so forth.

By definition, $TS_m = S_m^2$. The set $\left[TS_0 \ TS_1 \ TS_2 \ ... \ TS_m \right]$ is by implication a proxy map of the *temporal partial sums of squares* stream in the phylogenetic process. After division of the elements by the degrees of freedom, as in

$$V_i = \frac{TS_i}{k_i - 1}$$

we have the temporal partial variance stream $[V_0\ V_1\ V_2\ ...\ V_m]$. We take this as a proxy mapping of the total signal stream through the nodes of the phylogenetic tree. We note for the sake of completeness that in the sample of n hierarchical relevés, k_i and n_{ij} remain unchanged. In other words, the design of the dendrogram is rigidly fixed to the sample's species list. The temporal partial sum of squares stream is $[TS_0\ TS_1\ TS_2\ ...\ TS_m]$. This is the *characteristic profile* of the m-level dendrogram of the metacommunity. The matrix $[TS_{izy}], i = 0,1,2,...,m; z \neq y$ is a characteristic profile too, but in this the elements are *temporal partial sums of products* specific to the hierarchical relevés of two metacommunities z and y. The defining equations are:

$$S_{izy} = \sum_{j=1}^{k_i} n_{ij} \left(\frac{X_{ijz}}{n_{ij}} - \overline{X}_{0z} \right) \left(\frac{X_{ijy}}{n_{ij}} - \overline{X}_{0y} \right) ,$$

$$TS_{izy} = S_{izy} - S_{i+1zy}$$

and

$$r_{izy} = \frac{TS_{izy}}{\sqrt{TS_{iz}TS_{iy}}}$$

The latter is a standardised co-linearity scalar known as the *partial correlation coefficient*. As defined, the metacommunities z and y are compared on dendrogram level i. The values of r range from -1 to 1. A zero value indicates no relationship, positive r indicates co-linearity, a form of convergence or parallelism, and negative r signifies proportionately intense divergence, that is, the proportional lack of positive co-linearity. The *m* sets of *q(q-1)/2* distinct correlation values can be presented for further analysis by multivariate methods (Orlóci 2010).

SIGNAL ISOLATION

As the first step in signal analysis, we pool all hierarchical relevés on the gradient into a single metacommunity, with its own hierarchical relevé. We compute the *temporal partial variance stream* for this the same way as before:

$$
\mathbf{V}_{pooled} = \begin{bmatrix} V_{0pooled} \\ V_{1pooled} \\ \dots \\ V_{mpooled} \end{bmatrix}
$$

This is the best *temporal partial variance stream* we can get in the sample.

Forging on with the calculations, we compute an *m*-valued temporal partial variance stream for each of the meta-communities of the *q* gradient segments. We present these as columns in matrix $\mathbf{V}$ (see above). With this step, we have all ingredients -- the pooled temporal variance stream $\mathbf{V}_{pooled}$, the *m* spatial variance streams in the rows of matrix $\mathbf{V}$, and the gradient variable U --needed to proceed to the next step in signal isolation: computation of the gradient based slopes of the *m* spatial partial variance streams. For the ith of these, the gradient partial variance slope is

$$\alpha_i = abs \arc \tan \left(\frac{\sum\limits_{j=1}^{q}\left(V_{ij} - \overline{V_i}\right)\left(U_j - \overline{U}\right)}{\sum\limits_{j=1}^{q}\left(U_j - \overline{U}\right)^2} \right), \; i = 0,1, \; ..., \; m$$

Symbols $\overline{V_i}$ and $\overline{U}$ represent mean values.

The tn α_i quantity is best to obtain as a linear regression coefficient -- usual symbol **b**. Regression analysis, performed on each of the m rows in matrix **V** using U as the x variable, hands us as a bonus some useful statistics as seen in the results.

We consider further the m-valued vector of regression coefficients **b**. From this using the abs arc tan b transformation we obtain an m-valued angles vector α. Based on α, two sets of m quantities are defined:

$$PGS_i = V_{ipooled}(\cos\alpha_i), \; i = 0,1, \; ..., \; m$$

and

$$CES_i = V_{ipooled}(\sin\alpha_i), \; i = 0,1, \; ..., \; m$$

PGS_i is the phylogenetic signal's strength on level i of the dendrogram in the pooled metacommunity. Put it in another way, $\dfrac{100\ PGS_0}{PGS_0 + CES_0}\%$ of the diversity amplitude in the metacommunity is attributable to the splitting up of genera into species. Conversely, $\dfrac{100\ CES_0}{PGS_0 + CES_0}\%$ is the portion of the diversity amplitude attributable to environmental mediation by variable U. In still more general terms, the intensity of PGS on hierarchical level i is $\dfrac{100\ PGS_i}{PGS_i + CES_i}\%$. The one complement of

Signal isolation

this $\dfrac{100\ CES_i}{PGS_i + CES_i}$% is the strength of *CES* referenced to variable

U. In terms of grand totals, the proportions are

$$SSO = \frac{100 \sum\limits_{i=0}^{m} PGS_i}{\sum\limits_{i=0}^{m} PGS_i + \sum\limits_{i=0}^{m} CES_i}\%\ \text{for self-organisation, and}$$

and

$$EM = \frac{100 \sum\limits_{i=0}^{m} CES_i}{\sum\limits_{i=0}^{m} PGS_i + \sum\limits_{i=0}^{m} CES_i}\%\ \text{for environmental mediation.}$$

In summary, SSO scales the level of self organization and EM scales the strength of environmental mediation in the gradient-wide metacommunity. We may take the error terms associated with **b** as an estimate of e (see regression analysis in Orlóci 2010).

EXAMPLE

Survey site

We use M. Mihály's data set (Table 1) collected in the course of a transect survey of the floodplain on the inside of a major meander of the Coquihalla River floodplain near Hope, British Columbia. At the time of the survey (July 7-8, 1976) a natural high forest covered the site, now it is occupied by a subdivision of Hope as can be seen on the Google map (Fig. 2)[5].

The map identifies the end points of the transect line by markers, but the site itself is no longer the same. The river flow was regulated, many of the trees cut down, and part of the site bulldozed to accommodate the dwellings. The photograph of a nearby site (Figure 2) depicts similar conditions as have existed at the transect site at the time of survey. The vegetation formation of the region is identified by Krajina (1959) as part of the Pacific Northwest's Coastal Western Hemlock Zone.

---

[5] http://maps.google.com/maps/ms?ie=UTF&msa=0&msid=21662637730 9845313321.00049bf3254ab94ea225e

A common characteristic of floodplains is the presence of more or less flat benches (levels, terraces) with a slight sloping toward the adjacent higher terrain. The benches mark the joint effect of sedimentation and erosion by the flood water. The benches on the Coquihala transect are lined up on the across the transect with the elevation increasing as distance from the river increases. The vegetation cover of the benches is remarkably different at different elevations. This is an indication of different overflow frequency and duration, and the quality and quantity of the sediment load carried by the floods.

The surface morphology of the floodplain in the study site was structured in its natural state by two active benches and one fossil terrace. The average elevation of these was 4.2 m, 5.4 m, and 10.8 m above the water level of the river when the survey began (see date above).

Data set

The original survey contains 45 sample plots, 10m x 10m quare each. Cover/abundance values were estimated for nearly 100 species within the sample plots. A composite soil sample was taken from the top 20 cm of the soil profile for chemical analysis within each plot.

The two benches and one fossil terrace delineates three metacommunities (a,b,c in Table 1), described by average species cover/abundance and by gradient variables including elevation, soil nitrate nitrogen, ammoniac nitrogen, phosphorus, calcium, potassium and pH.

We use the records of the 40 most abundant species and the first three environmental variables in the analyses. We opted for such a reduction of variables to avoid overwhelming the example with information we did not need to make our point: the environmental signal *(CES)* is strong, but the phylogenetic signal *(PGS)* is dominant.

Table 1. Data set from the Coquihalla transect. A brief description of site, sampling design and data ownership appears in the main text. Items in table headings: # -- sequence number as in M. Mihály's original records; Code – identifies species belonging among the given categories; *FT* – functional type (life-form) code and average *CA* as weight; *ET* – ecological (flood duration) type code and average *CA* as weight; a,b,c – benches low to high, average heights 4.2 m, 5.4 m, 10.8 m; Poole sample (t) grand mean *CA* – calculated for entire transect based on 45 sample plots; Soil variables – weighted mean values *CA x NN kg/ha* and *CA x AN kg/ha; NN* – nitrate nitrogen, *AN* – ammoniac nitrogen. The total number of species in the analysis is 40 and the sample size is 45 plots. The terraces a, b, c are represented by 14, 20, 11 plots in that order. Plant identification followed standard field manuals.[6] The soil chemical analysis was performed by personnel in the Department of Agriculture, Guelph, Ontario.

#	Species	Class or higher	Code	Order	Code	Family	Code	Genus	Code
2	Acer macrophyllum	Eudicots	3	Sapindales	16	Sapindaceae	1	Acer	1
25	Galim triflorum	Asterids	1	Gentianales	10	Rubiaceae	2	Galim	11
16	Claytonia sibirica	Magnoliopsida	4	Caryophyllales	4	Portulacaceae	5	Claytonia	6
39	Osmorhiza chilensis	Asterids	1	Apiales	1	Apiaceae	21	Osmorhiza	21
37	Mnium insigne	Bryopsida	2	Eubryaales	9	Mniaceae	9	Mnium	20
3	Acer macrophyllum	Eudicots	3	Sapindales	16	Sapindaceae	1	Acer	1
15	Circaea alpina	Magnoliopsida	4	Myrtales	12	Onagraceae	7	Circaea	5
20	Dicentra formosa	Magnoliopsida	4	Ranunculales	14	Fumariaceae	12	Dicentra	8
56	Smilacina stellata	Monocots	5	Asparagales	2	Ruscaceae	21	Smilacina	29
60	Symphoricarpos albus	Asterids	1	Dipsacales	7	Caprifoliaceae	18	Symphoricarpos	31
43	Polystichum munitum	Pteridopsida	6	Dennstaedtiales	6	Dennstaedtiaceae	14	Polystichum	23
22	Disporum hookerii	Monocots	5	Liliales	11	Colchicaceae	16	Disporum	9

[6] C.L.A. Cronquist, M. Owenbey and J.W. Thompson. 1955-1959. Vascular plants of the Pacific Norhwest. Univ. of Washington Press, Seattle, Washington. Grout, A.J. 1928-1940. Moss flora of North America north of Mexico. Newfane, Vermont.

Example. Data set

30	Lactuca canadensis	Asterids	1	Asterales	3	Asteraceae	20	Lactuca	16
37	Mnium spinulosum	Bryopsida	2	Eubryales	9	Mniaceae	9	Mnium	20
67	Tsuga heterophylla	Pinopsida	6	Pinales	13	Pinaceae	6	Tsuga	34
17	Clintonia uniflora	Monocots	5	Liliales	11	Liliaceae	10	Clintonia	7
54	Rubus spectabilis	Magnoliopsida	4	Rosales	15	Rosaceae	3	Rubus	28
27	Goodyera oblongifolia	Monocots	5	Asparagales	2	Orchidaceae	7	Goodyera	13
65	Trientalis latifolia	Eudicots	3	Ericales	8	Myrsinaceae	8	Trientalis	33
45	Pteridium aquilinum	Pteridopsida	6	Dennstaedtiales	6	Dennstaedtiaceae	14	Pteridium	25
49	Rhytidiadelphus loreus	Bryopsida	2	Eubryales	9	Hypnaceae	11	Rhytidiadelphus	26
34	Mahonia nervosa	Magnoliopsida	4	Ranunculales	14	Berberidaceae	19	Mahonia	19
24	Eurhynchium oreganum	Bryopsida	2	Eubryales	9	Hypnaceae	11	Eurhynchium	10
8	Amelanchier florida	Magnoliopsida	4	Rosales	15	Rosaceae	3	Amelanchier	3
70	Vaccinium parviflorum	Asterids	1	Ericales	8	Ericaceae	13	Vaccinium	35
28	Holodiscus discolor	Magnoliopsida	4	Rosales	15	Rosaceae	3	Holodiscus	12
71	Vaccinium membranaceum	Asterids	1	Ericales	8	Ericaceae	13	Vaccinium	35
57	Spiraea douglasii	Magnoliopsida	4	Rosales	15	Rosaceae	3	Spiraea	30
33	Lonicera ciliata	Asterids	1	Dipsacales	7	Caprifoliaceae	18	Lonicera	18
52	Rosa gymnocarpa	Magnoliopsida	4	Rosales	15	Rosaceae	3	Rosa	27
62	Thuja plicata (shrub)	Pinopsida	6	Pinales	13	Cupressaceae	15	Thuja	32
13	Chimaphila umbellata	Asterids	1	Ericales	8	Pyrolaceae	4	Chimaphylla	4
32	Linnaea borealis	Asterids	1	Dipsacales	7	Caprifoliaceae	18	Linnaea	17
29	Hylocomium splendens	Bryopsida	2	Eubryales	9	Hypnaceae	11	Hylocomium	15
26	Gaultheria shallon	Eudicots	3	Ericales	8	Ericaceae	13	Gaultheria	12
40	Pachistima myrsinites	Eudicots	3	Celastrales	5	Celastraceae	17	Pachistima	22
43	Pseudotsuga menziesii	Pinopsida	6	Pinales	13	Pinaceae	6	Pseudotsuga	24
3	Acer macrophyllum (shrub)	Eudicots	3	Sapindales	16	Sapindaceae	1	Acer	1
6	Achlys triphylla	Magnoliopsida	4	Ranunculales	14	Berberidaceae	19	Achlys	2
63	Thuja plicata	Pinopsida	6	Pinales	13	Cupressaceae	15	Thuja	32

Table 1 continued

#	Species	FT C ode	FT Weight CA	ET C ode	ET Weight CA	Species mean CA				Soil variables, weighted means	
						Bench "a"	Bench "b"	Bench "c"	Pooled sample	CA$_X$ NN kg/ha	CA$_X$ AC kg/ha
2	Acer macrophyllum	1	3.26	1	1.36	1.86	0.30	0.00	0.72	14.23	40.35
25	Galim triflorum	3	0.67	1	1.36	1.29	0.90	0.00	0.73	12.24	30.26
16	Claytonia sibirica	5	0.75	1	1.36	1.86	0.00	0.00	0.62	15.24	43.71
39	Osmorhiza chil.	3	1.57	1	1.36	1.79	0.50	0.00	0.76	13.65	38.11
37	Mnium insigne	4	0.12	1	1.36	1.07	0.10	0.00	0.39	14.62	41.47
3	Acer macrophyllum	1	3.26	1	1.36	1.64	0.15	0.45	0.75	12.81	33.63
15	Circaea alpina	3	1.57	1	1.36	2.31	0.00	0.00	0.77	15.24	44.83
20	Dicentra formosa	5	1.31	1	1.36	3.36	0.25	0.00	1.20	14.74	42.59
56	Smilacina stellata	5	1.31	1	1.36	2.29	0.70	0.00	1.00	13.54	35.87
60	Symphoricarpos a.	2	1.91	1	1.36	5.00	0.90	0.36	2.09	13.64	36.99
43	Polystichum m.	4	0.12	1	1.36	7.50	2.35	0.09	3.31	13.43	34.75
22	Disporum hookerii	5	1.31	1	1.36	4.93	2.50	0.00	2.48	12.79	32.51

30	Lactuca canadensis	3	1.57	1	1.36	3.93	2.70	0.00	2.21	12.28	31.38
37	Mnium spinulosum	4	0.12	1	1.36	4.93	0.85	0.18	1.99	13.91	39.23
67	Tsuga heterophylla	1	3.26	2	1.55	0.00	3.05	0.00	1.02	7.96	19.05
17	Clintonia uniflora	5	1.31	2	1.55	0.00	4.10	0.27	1.46	7.81	16.81
54	Rubus spectabilis	2	1.91	2	1.55	0.57	2.15	1.45	1.39	8.14	21.30
27	Goodyera o.	5	1.31	2	1.55	0.00	1.95	1.36	1.10	6.99	13.45
65	Trientalis latifolia	3	1.57	2	1.55	0.50	1.90	1.73	1.38	7.86	17.93
45	Pteridium aqui.	4	0.12	2	1.55	0.00	2.70	0.00	0.90	7.96	20.18
49	Rhytidiadelphus l.	4	0.12	2	1.55	1.79	2.35	3.09	2.41	8.75	24.66
34	Mahonia nervosa	2	1.91	2	1.55	2.64	6.60	3.91	4.38	8.72	23.54
24	Eurhynchium o.	4	0.12	2	1.55	3.00	4.95	4.64	4.20	8.83	25.78
8	Amelanchier florida	2	1.41	3	1.84	0.21	0.40	1.45	0.69	7.06	14.57
70	Vaccinium p.	2	1.91	3	1.84	0.00	0.25	2.73	0.99	5.80	7.85
28	Holodiscus discolor	2	1.91	3	1.84	0.00	0.00	1.18	0.39	5.60	1.12
71	Vaccinium m.	2	1.91	3	1.84	0.00	0.00	2.09	0.70	5.60	2.24
57	Spiraea douglasii	2	1.91	3	1.84	0.00	0.00	1.91	0.64	5.60	4.48
33	Lonicera ciliata	2	1.91	3	1.84	0.00	0.25	1.55	0.60	5.93	8.97
52	Rosa gymnocarpa	2	1.91	3	1.84	0.14	0.40	1.36	0.64	6.82	12.33
62	Thuja plic. (shrub)	2	1.91	3	1.84	0.00	0.00	0.73	0.24	5.60	3.36
13	Chimaphila u.	4	0.12	3	1.84	0.00	0.20	3.09	1.10	5.75	6.73
32	Linnaea borealis	4	0.12	3	1.84	0.00	0.10	3.00	1.03	5.68	5.60
29	Hylocomium s.	4	0.12	3	1.84	1.64	6.15	6.64	4.81	7.71	15.69
26	Gaultheria shallon	2	1.91	3	1.84	0.00	2.15	7.45	3.20	6.13	10.09
40	Pachistima m.	2	1.91	3	1.84	0.29	3.25	6.55	3.36	6.64	11.21
43	Pseudotsuga m.	1	3.26	4	5.16	8.15	6.05	8.00	7.40	9.79	26.90
3	Acer macrophyllum	1	3.26	4	5.16	7.57	5.25	2.64	5.15	11.13	29.14
6	Achlys triphylla	3	1.57	4	5.16	4.50	4.70	1.55	3.58	10.67	28.02
63	Thuja plicata	1	3.26	4	5.16	2.71	6.10	4.73	4.51	8.60	22.42

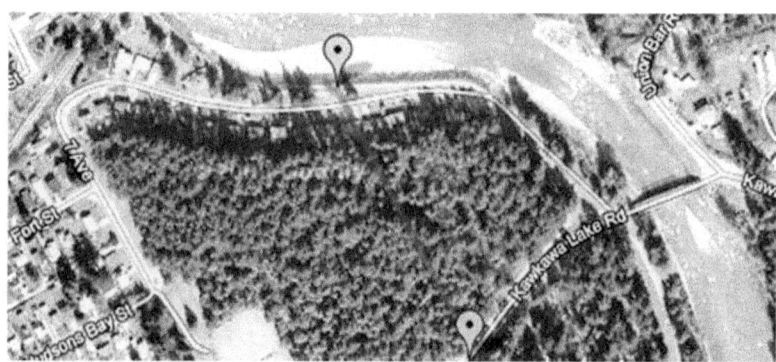

Figure 2. The Coquihalla River floodplain survey site at Hope, British Columbia. The transect site and sampling details are described in the main text and in the caption of Table 1. The flow in the stream can be torrential and several meters high-

er than on day of sampling. The flow from October to March is a mere average 2 m^3/s. This can reach 12 m^3/s in May and June following runoff after warm days or rain. The river is fed by runoff from a 740 km^2 watershed in the Cascade Mountains. The highest peak in the photograph is nearly 2000 m.

Signal numerics

All computed results are presented in the appendices. For ease of access the partial variance streams (V matrix, V_{pooled} vector) are reproduced in Table 2, selected results from regression analyses in Table 3, and from correlation analysis in Table 4. Empirical probabilities (Figure 3) for the correlations were determined in a Monte Carlo experiment under the assumption of zero expected correlation.

We recall that the sample size of the grand pooled meta-community is 45, and for the metacommunities of the a, b, c transect segments 14, 20, 11 in that order. The number of elements in the baseline data is uniformly 40 (the number of species in the analysis). The degrees of freedom (DF) are a function of the hierarchical levels:

Level in hierarchy	Number of records in cumulants	DF	Partial DF
4	6	5	5
3	17	16	11
2	24	23	7
1	35	34	11
0	40	39	5

Table 2. Spatial partial variance streams is in rows (hierarchical levels 0 to 4). Temporal partial variance streams of the floodplain level metacommunities are in columns a, b, c , the pooled metacommunities partial variance stream in last col-

umn *(t)*. Gradient variable *U* is floodplain elevation. See further explanations in the main text.

Levels	*V* matrix			*V*_{pooled} vector
	a	b	c	t
0 Species	6.74012	7.139820	2.437623	4.687
1 Genus	7.348533	1.586646	4.075429	2.585
2 Family	3.304415	6.150125	7.23172	2.709
3 Order	4.968971	3.435011	5.359622	1.519
4 Class	2.767431	7.020941	5.937247	3.882
Variable *U m*	4.2	5.4	10.8	

Table 3. Regression analysis summarised. The y variable is the spatial partial variance stream (rows 0 to 4 in the table). The x variable is floodplain elevation *U.* Columns in upper table: temporal partial variance streams in the metacommunities of the three transect segments (a, b, c), regression coefficients (*b* = tan α) , the standard deviation (SD) of tan α, coefficient of determination *(R^2)*, spatial specific variance slope *(α^o)*, probability of a more extreme α obtained by chance *(P > α)*, strength of the environmental effect (variable *U)* in $(1-P^2)^{1/2}$ terms. Columns in lower table: temporal specific variances of the pooled metacommunity (t), the phylogenetic signal *(PGS and PGS %)* and the environmental signal *(CES and CES%)*.

Levels	a	b	c	b= tan α	e=SD	R^2	α^o	P>α^o	$(1 - P^2)^{1/2}$
0 Species	6.74	7.14	2.43	-0.72	0.182	0.92	35.7	0.059	0.998
1 Genus	7.35	1.59	4.07	-0.20	0.796	0.06	11.5	0.822	0.569
2 Family	3.30	6.15	7.23	0.474	0.327	0.68	25.4	0.286	0.958
3 Order	4.97	3.43	5.36	0.150	0.247	0.27	8.53	0.606	0.796
4 Class	2.77	7.02	5.93	0.27	0.56	0.18	15.22	0.67	0.735

Example. Signal numerics

	$t*$	$PGS = t \times \cos \alpha*$	$CES = t \times \sin \alpha$	% PGS	% CES
0 Spp.	4.6868	3.8056	2.7355	65.9338	34.0662
1 Genus	2.5847	2.5329	0.5149	96.0310	3.9690
2 Fam.	2.7087	2.4473	1.1608	81.6353	18.3647
3 Order	1.5190	1.5022	0.2255	97.7966	2.2034
4 Class	3.8825	3.7463	1.0190	93.1113	6.8887
Average	3.0763	2.8069	1.1311	**86.9016**	**13.0984**

* Pythagorean theorem applied: e.g. $4.686753^2 = 3.805626^2 + 2.735484^2$

Table 4. Correlations analysis of the t, FT, ET, NN, AN variables on floodplain elevation U identified in caption of Table 1. Columns: temporal partial correlation streams in top portion, probabilities of an at least as extreme absolute correlation as the observed in middle portion, correlation strengths in bottom portion. See probability equation in Figure 3 and the explanations in the main text.

Partial correlations

Level	txFT	TxET	txNN	txAA
0 Species	0.624	0.970	-0.081	-0.341
1 Genus	0.066	0.490	0.434	-0.143
2 Family	0.266	0.613	-0.557	0.443
3 Order	0.253	0.604	0.066	-0.144
4 Class	0.078	0.573	0.221	-0.517

FTxET	FTxNN	FTxAA	ETxNN	ETxAA	NNxAA
0.603	0.683	-0.900	-0.091	-0.324	-0.908
-0.164	0.350	-0.399	0.107	0.116	-0.883
0.178	-0.035	-0.063	-0.623	0.554	-0.939
0.818	-0.083	0.146	-0.100	0.130	-0.983
0.684	-0.471	0.246	-0.474	0.078	-0.744

The probability of an at least as extreme correlation as the observed occurring by chance, symbolically $P(r_{RND} \geq r)$ (see the main text and Figure 3).

Level	txFT	TxET	txNN	txAA
0 Species	0.486	0.109	0.872	0.623
1 Genus	0.895	0.550	0.575	0.790
2 Family	0.672	0.492	0.520	0.571
3 Order	0.682	0.497	0.895	0.789
4 Class	0.877	0.512	0.709	0.538

FTxET	FTxNN	FTxAA	ETxNN	ETxAA	NNxAA
0.766	0.618	0.592	0.835	0.823	0.251
0.751	0.946	0.900	0.487	0.522	0.164
0.332	0.869	0.786	0.845	0.806	0.084
0.449	0.559	0.688	0.557	0.877	0.404

Correlation strength expressed by the coherence coefficient $R = \sqrt{1 - P^2}$.

Level	txFT	TxET	txNN	txAA
0 Species	0.874	0.994	0.489	0.782
1 Genus	0.446	0.835	0.818	0.613
2 Family	0.740	0.871	0.854	0.821
3 Order	0.731	0.868	0.446	0.615
4 Class	0.481	0.859	0.705	0.843
Averag	**0.655**	**0.885**	**0.663**	**0.735**

FTxET	FTxNN	FTxAA	ETxNN	ETxAA	NNxAA
0.868	0.893	0.974	0.514	0.774	0.977
0.643	0.786	0.806	0.550	0.567	0.968
0.661	0.325	0.437	0.874	0.853	0.986
0.943	0.495	0.618	0.535	0.593	0.996
0.893	0.829	0.726	0.830	0.481	0.915
0.802	0.666	0.712	0.661	0.654	0.969

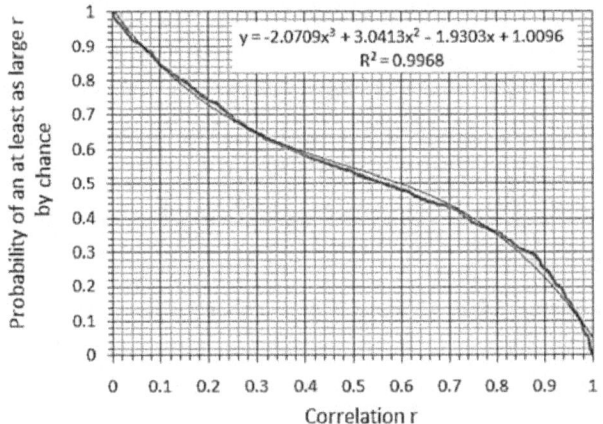

Figure 3. Empirical probabilities of a random *r* being at least as large or larger than the probability points (horizontal axis). See explanations in the main text and in Table 4. Empirical probabilities were determined in Monte Carlo experiments (see Orlóci 2010) under the assumption zero expectation. The smaller is this kind of probability the more significant is the *r* actually observed. Equation: y probability; x correlation.

Considering Table 2, the temporal partial variances are laid out in columns a, b, c for the metacommunities of the 3 transect segments. The rows (0 to 4) contain the spatial variance streams. We subjected each of the spatial partial variance streams to regression analysis using floodplain elevation *(U)* as the *x* variable. The regression results are presented in Table 3.

Our focus is in each case on the partial variance slope *(α⁰)*. This measures the strength of environmental mediation *(1-$P^2)^{1/2}$*. We also give the relative size of the isolated signals *PGS* and *CES*. From these values we conclude that on the Species (baseline) level and on the Family level the floodplain elevation effect is numerically high and in probability terms very robust. On the Genus level it is very low and weak in *TS*, on the Order level weak but quite robust, and on the Class level numerically weak but still quite robust. This already suggests that the relative importance of *CES* varies; strong on some levels of the dendrogram and weak on other levels. We can make a definitive statement (generalization) based on the isolates, *CES* from *PGS*. The sizes of these show the dominance of *PGS* (66%) to *CES* (44 %) on the Species level, and 87% to 14% on average through all taxonomic levels.

Continuing with the interpretation of results, we turn now to Table 4 and Figure 3. In these we can see the environmental effects on *TS* in partial correlation terms. The correlation values are based on 5 hierarchical relevés of the gradient wide metacommunity with base line variables *t, FT, ET, NN,* and *AN.*

We note at the outset that our interpretation of the strength of the correlation coefficient relies on two things: the magnitude of *r* and the probability of obtaining a correlation value equal to or greater than the observed *r* under the assumption that the expectation of *r* is zero, symbolically $P= P(r_{RND} \geq r)$. If absolute *r* is numerically strong then $R = \sqrt{1 - P^2}$ approaches the upper limit of its interval 0 to 1. In formulating *R*, we use Rajski's information theoretical considerations (Orlóci 2010). We call *R* the *coherence coefficient*. Note that in Rajski's scheme the lack of correlation has counterpart not in the one-complement *1- P*, but in *R*. In other words, *R* and *P* are considered orthogonal.

Dendrogram with baseline values in the tth column of Table 1 is the base dendrogram in correlation analysis. Looking at the numerics, the correlations suggest the following strong relationships:

a. The species based hierarchy *(t)* has high coherence with the ecological type *(ET)* hierarchy on all hierarchical levels. This suggests that the species represent strong ecological traits as Pillar and Duarte (2010) concluded. The *t* hierarchy scores high coherence also with the functional type *(FT)* hierarchy on the hierarchical levels except Genus and Class.

c. Soil nitrate nitrogen *(NN)* fails in similar terms as *FT* in achieving a high correlation, unlike the coherence of the *t* hierarchy and the ammoniac nitrogen *(AN)* whose correlation is strong.

d. The coherence of the *AN* and *NN* hierarchies is very high but negative. This should not be surprising since the humus type changes from the quickly decomposing kind on the functioning floodplain benches to deep raw humus on the fossil terrace.

DISCUSSION

Kerner's (1863) ideas on the assembly of species into environment specific communities and his assertion that plant communities develop *in situ* in the sense of undergoing compositional transitions, owing to a feedback mechanism which he discovered and we call facilitation, was truly the kernel of modern dynamic ecology. Kerner's work is seminal on the environmental mediation rule and starting point toward finding general rules of governance under which plant community assembly/disassembly can proceed as a non-random process. Now it is clear that to find general rules the search has to go beyond the Kernerian environmental mediation rule and scrutinise its dual process powered by interactions of the community elements (Stachowicz 2001, Tilman 2004). Ecologists call the dual process *self-organization* (Camazine et al. 2003).

Wilson (2009) makes these remarks:

"... Two basic kinds of community pattern can be envisaged, with different causes:

(a) Environmentally mediated patterns, i.e., correlations between species due to their shared or opposite responses to the physical environment. Ecologists have long tried 'to find out which species are commonly associated together upon similar habitats' (Warming, 1909). Modern methods allow more subtle questions to be examined, such as the shape of environmental responses (e.g., Bio et al., 1998), the niche widths (e.g., Diaz et al., 1994), and how repeatable the associations of species are (e.g., Wilson et al., 1996c). However, the simple existence of environmentally mediated patterns is now too obvious to need demonstrating; Warming (1909) described it as 'this easy task'.

(b) Assembly rules, i.e., patterns due to interactions between species, such as competition. These patterns, when we can find them, are fascinating evidence that competition, allelopathy, facilitation, mutualism, and all the other biotic interactions that we know about in theory, actually affect communities in the real world.

Of course, to make this distinction, it has to be known what processes have caused each pattern – physical environment or biotic interactions – but that is our task as community ecologists. Both types of process may occur."

Wilson's point (a) confirms environmental mediation by mechanisms which use facilitation. His point (b) addresses self organization by biological interactions.

Interesting to note how Wilson perpetuates a misconception adopted widely in the English language literature, when he points to Warming (1909) as his historic reference rather than mentioning Kerner (1863). Kerner does in fact discuss in *"Das Pflanzenlebel der Donau Lander"* the process in a feedback point process context of facilitation, narrating how plant species can change the environment eventually to their own demise in the site.

Based on our results, it can be seen quite clearly that environmental mediation promotes the community's transience, and self-organization tends to promote the community's stability and permanence. Both processes are point processes and for that reason they are running in tandem.

Actually measuring the level of environmental mediation and self-organization takes us a good step further. We use for this a methodology which is centred on signal superposition $TS=PGS+CES+e$.

The isolation and scaling of signals is a topic discussed at length in many publications (see Pillar and Duarte 2010 and references therein) with varying success (Revell et al. 2008). Pillar and Duarte (2010) see the phylogenetic signal's role in community assembly in the manner of species being sorted by their environmental traits. This is an important assembly rule which our results support by showing the dominance of *PGS* related self organization over *CES* related transience. Formally recognizing this is a strong statement contrary to the spreading arguments which give dominance to the rule of chance.

Why is this surge of interest in the isolation of the signals? An answer to this question is the ecologists' intention to discover and then incorporate provisions for the phylogenetic process into the main corpus of governance principles in community assembly (Orlóci and He 2009). Pillar and Duarte (2010) mention several benefits to be gained from signal research such as the comparison of metacommunities, the discovery of new community assembly rules, the identification of plant species traits most symptomatic for the phylogenetic signal, the illumination of trait-filtering along ecological gradients, and not the least, the addressing of questions regarding phylogenetic niche conservatism.

We go a step further based on arguments we have already given. Those are strong reason to stand behind the propositions we listed at the beginning of the main text. Regarding the example, some of our Coquihalla results are simply astonishing. The many fold dominance the phylogenetic signal (SSO = 86.6%) over the environmental signal (EM = 13.1%) is nothing less than such. We conclude on that basis that the importance of self-organization overwhelms the importance of environmental mediation. The forgoing implies that community transience is not as strong a tendency as competition based self-organization which may counteract the tendency to community transience. The forgoing also implies that the moderation of the diversity amplitude in metacommunities may be far more a phylogenetic functionality than an environment mediated functionality.

QUESTIONS AND ANSWERS

It is fitting to end the book with questions that we can expect from the more thoughtful reader:

Q1. All considered, are the conclusions based on the numerical results, regarding the relative importance of the signals, general in relevance?

2. How much what has been concluded depend on the surveyed forest's mature natural state?

3. Should we expect the conclusions to change if, instead, we had used data collected on a site actively recovering from disturbance?

None of the questions appear to cast any doubt upon the validity of the 4 major propositions and the two corollaries that we accept as facts. Their validity does not derive from nor can be validated by a simple example. The conclusions based on the numerics are different. The SSO/EM fraction, suggesting a close to 7 folds proportional importance for *PGS* over *CES* (see Table 3), is a site related fact. We do not know, because we did not study, how the values of SSO/EM evolve secondary succession. What we suspect is that in the initial explosive phase (see Orlóci 2009, Orlóci and He 2009) the *spatial partial variance angle* may reach high values, in which case the SSO/EM ratio is reduced. But the *spatial partial variance angle* must be greater than 45 degrees for the value of SSO/EM to drop below 1 (see the table below).

$\alpha°$	PGS	CES	SSO/EM
0.00	1.00	0.00	--
10.00	0.98	0.17	5.67
20.00	0.94	0.34	2.75
30.00	0.87	0.50	1.73
40.00	0.77	0.64	1.19
45.00	0.71	0.71	1.00
50.00	0.64	0.77	0.84
60.00	0.50	0.87	0.58
70.00	0.34	0.94	0.36
80.00	0.17	0.98	0.18
90.00	0.00	1.00	0.00

None of the angles in Table 3 come even close to the 45 degree mark. It is easily verified that the ratio is indeterminate when the angle is zero and zero when the angle is 90 degrees. But such extremes are not expected in Nature.

It is important when readers try to recalculate partial results they must be aware of our assumption of a unit temporal partial variance in the pooled metacommunity. Extra attention needed in the calculations to use radians when the calculator requires the input of radians. The conversion is $1° = 3.141593/180 = 0.017453294$ radians.

REFERENCES

Bentham, G. and J.D. Hooker. 1862–1883. Genera plantarum ad exemplaria imprimis in herbariis kewensibus servata definita. 3 volumes. Biodiversity Heritage Library, http://www.biodiversitylibrary.org/item/14680

Bio, A.M.F., Alkemande, R., and A. Barendregt. 1998. Determining alternative models for vegetation response analysis: a non-parametric approach. Journal of Vegetation Science 9: 5-16.

Camazine, S., Deneubourg, J.L., Franks, N.R. Sneyd, J., Theraulaz G. and E. Bonabeau. 2003. Self-Organization in Biological Systems, Princeton University Press.

Diaz, S., Acosta, A., and M. Cabido. 1994. Grazing and the phenology of flowering and fruiting in a montane grassland in Argentina – a niche approach. Oikos 70: 287-295.

Dobzhansky, T. 1937. Genetics and the Origin of Species. Columbia University Press.

Felsenstein, J. 2004. Inferring Phylogenies. Sinauer Associates, Sunderland, MA.

Greig-Smith, P. 1952. The use of random and contiguous quadrats in the study of the structure of plant communities. Annals of Botany 16: 293-316.

Hennig 1966. Phylogenetic systematics (tr. D. Dwight Davis and Rainer Zangerl). Urbana, IL: Univ. of Illinois Press. (Reprinted 1979 and 1999).

Huxley 1942. Evolution: the modern synthesis. The MIT Press.

Kerner von Marilaun, A. 1863. Das Pflanzenleben der Danauländer. Innbruck, Wagner.

Krajina, V.J. 1959. Bioclimatic Zones in British Columbia. UBC Botanical Series #1, Vancouver, B.C.

Mayr, E. and W.B. Provine (eds). 1998. The Evolutionary Synthesis: Perspectives on the Unification of Biology. Harvard University Press.

Mayr, E. 2002. What evolution is. Weidenfeld & Nicolson, London.

Orlóci, L. 1971. An information theory model for pattern analysis. Journal of Ecology 59:343-349.

Orlóci, L. 1991. On character-based community analysis: choice, arrangement, comparison. In: Feoli, E. and L. Orlóci (eds.), Computer Assisted Vegetation Analysis, pp. 81-93. Kluwer Academic Publishers, London.

Orlóci, L. 2006. Diversity partitions in 3-way sorting: functions, Venn diagram mappings, typical additive series, and examples. Community Ecology 7:253-259.

Orlóci, L. 2009. Multi-scale trajectory analysis: powerful conceptual tool for understanding ecological change. Frontiers of Biology in China 4: 158-179

Orlóci, L. 2010. Statistical Ecology: a reasoned approach. Scada Publishing. Internet Edition https://www.createspace.com/3476529

Orlóci, L. and M. Orlóci. 1985. Comparison of communities without the use of species: model and example. Ann. Bot. (Roma) 43:275-285.

Orlóci, L.and K.S. He. 2009. On Governance in the long-term vegetation process. How to discover the rules? Frontiers of Biology in China 4: 557-568.

Pillar, V.D. and L.S. Duarte. 2010. A framework for meta-community analysis of phylogenetic structure. Ecology Letters 13: 587–596.

Pillar, De Patta V. and L. Orlóci. 1993. Character-based Vegetation Analysis: the Theory and an Application Program. Ecological Computations Series (ECS): Vol. 5. SPB Academic Publishing bv, The Hague, The Netherlands.

Podani, J. 2003. The Evolution and Systematics of Terrestrial Plants. In Magyar. Elte Ötvös Kiadó, Budapest.

Podani, J. 2010. Taxonomy in Evolutionary Perspective. An essay on the relationships between taxonomy and evolutionary theory. Synbiologia Hungarica 6:1-42.

Rényi, A. 1961. On measures of entropy and information. In: J. Neyman (ed.), Proceedings of the 4th Berkeley Symposium on Mathematical Statistics and Probability, pp. 547-561. University of California Press, Berkeley.

Revell, L.J., Harmon, L.J. and D.C. Collar. 2008. Phylogenetic Signal, Evolutionary Process, and Rate. Oxford Journals, Life Sciences, Systematic Biology Volume57, Issue 4, Pp. 591-601.

Schuh, R.T. and A.V.Z. Brower. 2009. Biological Systematics: Principles and Applications. (2nd edn.) Cornell University Press .

Singh, G. 2004. Plant Systematics: An Integrated Approach. Science Publishers, Enfield, NH. ISBN 978-1-57808-351-0

Stachowicz, J.J. 2001. Mutualism, facilitation, and the structure of ecological communities. BioScience 51: 235-246.

Stebbins, G. L. 1950. Variation and Evolution in Plants. Columbia University Press, New York.

Sukopp, H. 1987. On the history of plant geography and plant ecology in Berlin. Englera 7: 85-103.

Tilman, D. 2004. Niche tradeoffs, neutrality, and community structure: A stochastic theory of resource competition, invasion, and community assembly. PNAS July 27, 2004 vol. 101, no. 30 10854-10861.

Warming, E. with M. Vahl. 1909. Oecology of Plants - an introduction to the study of plant-communities translated by P. Groom and I. B. Balfour. Clarendon Press, Oxford. ---- Original: Warming, E. 1895. *Plantesamfund* - *Grundtræk af den økologiske Plante-geografi*. P.G. Philipsens Forlag, Kjøbenhavn.

Wildi, O. and M Schütz. 2000. Reconstruction of a 405 yr. recovery process from pasture to forest Community Ecology 1: 25-32.

Wilson, J.B. 2009. Assembly rules in plant communities, pp.130-164. In: E. Weher and P. Keddy (eds.), Ecological Assembly Rules, Cambridge Books Online.

http://ebooks.cambridge.org/chapter.jsf?bid=CBO9780 511542237&cid=CBO9780511542237A013

Wilson, J.B., Ulman, I. and P. Bannister. 1996. Do species assemblages recur? Journal of Ecology 84: 471-474.

INDEX

APPENDICES

Appendix 1. Numerical values in this table are partial sums of squares in columns a, b, c, and t, and products in columns ab, ac, at, bc, bt, ct in Part I; partial variance and covariance values in Part II; and partial product moment correlations in Part III. The individual columns are temporal streams which map the total signal in the taxonomic dendrogram. The numerical results are based on the three metacommunities for which the hierarchical relevés are given in Table 1. Symbols a, b, and c are labels for the flood plain levels (3 transect segments and metacommunities occupying the segments) and for the pooled metacommunity of the entire transect (t). The basic data on which these results are based are in Table 1.

Part I   Temporal partial sum of squares and product streams

Level	a	ab	ac	at	b	bc	bt	c	ct	t
0 Species	33.701	29.183	15.031	25.979	35.699	20.361	28.425	12.188	15.869	23.434
1 Genus	80.834	12.666	23.765	39.147	17.453	19.725	16.631	44.830	29.441	28.432
2 Family	23.131	-4.720	-4.558	4.640	43.051	36.080	24.823	50.622	27.379	18.961
3 Order	54.659	23.440	-35.199	14.275	37.785	11.304	24.171	58.956	11.699	16.709
4 Class	13.837	20.954	11.147	15.315	35.105	15.822	23.974	29.686	18.909	19.412

Part II   Temporal partial variance and covariance streams

Appendices

Level	a	ab	ac	at	b	bc	bt	c	ct	t
0 Species	6.740	5.837	3.006	5.196	7.140	4.072	5.685	2.438	3.174	4.687
1 Genus	7.349	1.151	2.160	3.559	1.587	1.793	1.512	4.075	2.676	2.585
2 Family	3.304	-0.674	-0.651	0.663	6.150	5.154	3.546	7.232	3.911	2.709
3 Order	4.969	2.131	-3.200	1.298	3.435	1.028	2.197	5.360	1.064	1.519
4 Class	2.767	4.191	2.229	3.063	7.021	3.164	4.795	5.937	3.782	3.882

Part III   Temporal partial correlation streams

level	ab	ac	at	bc	bt	ct
0 Species	0.841	0.742	0.924	0.976	0.983	0.939
1 Genus	0.337	0.395	0.817	0.705	0.747	0.825
2 Family	-0.150	-0.133	0.222	0.773	0.869	0.884
3 Order	0.516	-0.620	0.472	0.239	0.962	0.373
4 Class	0.951	0.550	0.934	0.490	0.918	0.788

Appendix 2. Result of linear regression analysis. The spatial partial variance streams on which the analysis was performed are taken from Appendix 1.

x data: 4.2m, 5.4m, 10.8m
y data by dendrogram level :

0 Species	6.74012	7.139820	2.437623
1 Genus	7.348533	1.586646	4.075429
2 Family	3.304415	6.150125	7.23172
3 Order	4.968971	3.435011	5.359622
4 Class	2.767431	7.020941	5.937247

```
Hierarchical level 0
r^2 Coef Det    DF Adj r^2    Fit Std Err    F-val
0.9396348213  0.8792696427  0.9058709452  15.565841800
Parm  Value      Std Error    t-value      95% Confidence Limits    P>|t|
 a   10.32725653 1.344809123 7.679347469 -6.76016354 27.41467661 0.01654
 b   -0.71883366 0.182197367 -3.94535699 -3.03387071 1.596203398 0.05865
Hierarchical level I
r^2 Coef Det    DF Adj r^2    Fit Std Err    F-val
0.0611763926  0.0000000000  3.9598506641  0.0651628188
Parm  Value      Std Error    t-value      95% Confidence Limits    P>|t|
 a   5.719363854 5.878589358 0.972914335 -68.9751962 80.41392394 0.43322
 b   -0.20330802 0.796442771 -0.25527009 -10.3230729 9.916456910 0.82237
```

Hierarchical llevel 2
r^2 Coef Det   DF Adj r^2   Fit Std Err   F-val
0.6757177649 0.3514355298 1.6337135066 2.0837335254
Parm Value   Std Error   t-value   95% Confidence Limits   P>|t|
 a   2.336700874 2.425326521 0.963458262 -28.4799945 33.15339626 0.43697
 b   0.474321440 0.328587975 1.443514297 -3.70078466 4.649427536 0.28568
Hierarchical level 3
r^2 Coef Det   DF Adj r^2   Fit Std Err   F-val
0.2690170542 0.0000000000 1.2300725218 0.3680209719
Parm Value   Std Error   t-value   95% Confidence Limits   P>|t|
 a   3.567277126 1.826102005 1.953492804 -19.6355489 26.77010311 0.18998
 b   0.150086893 0.247403867 0.606647321 -2.99347730 3.293651088 0.60578
Hierarchical level 4
r^2 Coef Det   DF Adj r^2   Fit Std Err   F-val
0.1871937354 0.0000000000 2.8182138367 0.2303054781
Parm Value   Std Error   t-value   95% Confidence Limits   P>|t|
 a   3.392132709 4.183774409 0.810782891 -49.7677616 56.55202701 0.50263
 b   0.272020631 0.566825931 0.479901530 -6.93018570 7.474226967 0.67866

Appendix 3. Correlation analysis of variables t,FT,ET.NN,AN.

Symbols in column heading and associated data in Table 1: t - average cover abundance values of species in the pooled, gradient wide metacommunity; FT - functional type, cover abundance; ET - ecological type, cover abundance; NN - weighted nitrate nitrogen; AN - weighted ammonic nitrogen. The entries in main body of table are partial sum of squares and partial products in Part I, partial variances and co-variances in Part II, partial correlations and associated prob-abilities of an at least as extreme correlation by chance (Fig-ure 3) in Part III. Note, when we shift from the species based hierarchy (t) to another type, for example to the FT based hierarchy, the dendrogram remains the same as defined for species. Only the numerical values change in the baseline data and in the cumulants.

Part I

Level	txt	txFT	txET	txNN	txAN	FTxFT	FTxET	FTxNN	FTxAN
0 Species	23.434	2.882	18.273	-1.089	-30.662	0.911	2.241	1.802	-15.964
1 Genus	28.432	0.542	9.540	16.611	-29.053	2.394	-0.929	3.887	-23.430
2 Family	18.961	2.022	4.508	-21.222	73.051	3.043	0.523	-0.536	-4.142
3 Order	16.709	4.033	7.493	3.855	-45.973	15.188	9.669	-4.575	44.449

Appendices

4 Class	19.412	1.452	6.758	3.474	-60.683	17.975	7.765	-7.116	27.835

ETxET	ETxNN	ETxAN	NNxNN	NNxAN	ANxAN
15.138	-0.976	-23.426	7.631	-46.626	345.592
13.327	2.808	16.122	51.644	-241.114	1443.217
2.851	-9.206	35.407	76.666	-311.131	1433.339
9.207	-4.320	30.807	201.993	-1088.501	6064.259
7.159	-4.524	5.552	12.722	-70.735	710.285

Part II

Level 0	txt	txFT	txET	txNN	txAN	FTxFT	FTxET	FTxNN	FTxAN
Species 1	4.691	0.577	3.657	-0.218	-6.135	0.182	0.448	0.360	-3.193
Genus 2	2.585	0.049	0.867	1.510	-2.641	0.218	-0.084	0.353	-2.130
Family	2.709	0.289	0.644	-3.032	10.436	0.435	0.075	-0.077	-0.592
3 Order	1.519	0.367	0.681	0.350	-4.179	1.381	0.879	-0.416	4.041
4 Class	3.882	0.290	1.352	0.695	-12.137	3.595	1.553	-1.423	5.567

ETxET	ETxNN	ETxAN	NNxNN	NNxAN	ANxAN
3.03	-0.20	-4.685	1.526	-9.325	69.118
1.212	0.255	1.466	4.695	-21.919	131.202
0.407	-1.315	5.058	10.952	-44.447	204.763
0.837	-0.393	2.801	18.363	-98.955	551.296
1.432	-0.905	1.110	2.544	-14.147	142.057

Part III

Level	txFT	txET	txNN	txAN	FTxET	FTxNN	FTxAN	ETxNN	ETxAN	NVxAN
0 Species	0.624	0.970	-0.081	-0.341	0.603	0.683	-0.900	-0.091	-0.324	-0.908
1 Genus	0.066	0.490	0.434	-0.143	-0.164	0.350	-0.399	0.107	0.116	-0.883
2 Family	0.266	0.613	-0.557	0.443	0.178	-0.035	-0.063	-0.623	0.554	-0.939
3 Order	0.253	0.604	0.066	-0.144	0.818	-0.083	0.146	-0.100	0.130	-0.983
4 Class	0.078	0.573	0.221	-0.517	0.684	-0.471	0.246	-0.474	0.078	-0.744

Probability P of an at least as extreme correlation as the observed occurring by chance

Level	txFT	txET	txNN	txAN	FTxET	FTxNN	FTxAN	ETxNN	ETxAN	NVxAN
0 Species	0.486	0.109	0.872	0.623	0.497	0.450	0.226	0.858	0.633	0.214

1 Genus	0.895	0.550	0.575	0.790	0.766	0.618	0.592	0.835	0.823	0.251
2 Family	0.672	0.492	0.520	0.571	0.751	0.946	0.900	0.487	0.522	0.164
3 Order	0.682	0.497	0.895	0.789	0.332	0.869	0.786	0.845	0.806	0.084
4 Class	0.877	0.512	0.709	0.538	0.449	0.559	0.688	0.557	0.877	0.404

Strength of correlation expressed by the coherence coefficient $R = \sqrt{1 - P^2}$

Level	txFT	txET	txNN	txAN	FTxET	FTxNN	FTxAN	ETxNN	ETxAN	NVxAN
0 Species	0.874	0.994	0.489	0.782	0.868	0.893	0.974	0.514	0.774	0.977
1 Genus	0.446	0.835	0.818	0.613	0.643	0.786	0.806	0.550	0.567	0.968
2 Family	0.740	0.871	0.854	0.821	0.661	0.325	0.437	0.874	0.853	0.986
3 Order	0.731	0.868	0.446	0.615	0.943	0.495	0.618	0.535	0.593	0.996
4 Class	0.481	0.859	0.705	0.843	0.893	0.829	0.726	0.830	0.481	0.915

Reader notes

Reader notes

About the book

Community Ecology has a heavy core of assumptions which have yet to be interconnected on a common base to become parts of a general scientific theory. One of the assumptions is telling us that the assembly of a plant community is governed by rules other than a total reign of chance. The present book argues that the common base is community diversity and that on that basis assembly rules can be identified and expressed in quantitative terms. The proposed solution for quantification is signal theoretical, based on a stochastic model. The model incorporates independent channels for the historic phylogenetic signal, the current environmental signal, and the ubiquitous random effects. The first of the signal triplet modulates self-organisation and promotes community stability. The second mediates transience and fosters instability. And the third captures the level of aimlessness which is the level of signal garbling in the assembly process. What makes the solution optimal and why is solving the problem the way as it is proposed is important? Optimality comes with the triplet's encapsulation of the assembly rules. This is an asset, because it provides a basis for analytical isolation of the signals in quantitative terms, and through this, for establishing ecological operationality. The Book explains the theoretical implications of the solution and illustrates by data-based example the implementation's *modus operandi* in common practice.

In brief: what is inside the Book?

A novel, signal theoretical solution is sketched out for the ecological problem of how to identify and quantitatively express the assembly rules of plant communities. A case study for testing the solution leads to the astonishing conclusion that the phylogenetic signal outperforms the current environmental signal in intensity close to 7 to 1. This indicates high stability and low inclination to environment mediated transience in the community.